T0282646

CAMBRIDGE LIBRARY COLLECTION

Books of enduring scholarly value

Physical Sciences

From ancient times, humans have tried to understand the workings of the world around them. The roots of modern physical science go back to the very earliest mechanical devices such as levers and rollers, the mixing of paints and dyes, and the importance of the heavenly bodies in early religious observance and navigation. The physical sciences as we know them today began to emerge as independent academic subjects during the early modern period, in the work of Newton and other 'natural philosophers', and numerous sub-disciplines developed during the centuries that followed. This part of the Cambridge Library Collection is devoted to landmark publications in this area which will be of interest to historians of science concerned with individual scientists, particular discoveries, and advances in scientific method, or with the establishment and development of scientific institutions around the world.

Narrative of an Excursion to the Lake Amsanctus and to Mount Vultur in Apulia in 1834

This short but distinctive paper was published in 1835 by Charles Daubeny (1795–1867), who began his career as a physician but soon found his passion to be volcanos. At this time, Daubeny held chairs in chemistry and botany at Oxford. He had made many field trips to European volcanic regions between 1819 and 1825, was elected a Fellow of the Royal Society in 1822, and in 1826 published the first edition of his famous *Description of Active and Extinct Volcanos*, of which a later version also appears in this series. Here Daubeny describes a winter trip to the Apulia (Puglia) region in the south-east of Italy, rarely described by travel writers of his time, to visit Lake Amsanctus, famously mentioned by Virgil, and the extinct volcano Mount Vultur. Although Daubeny's overall focus is scientific, his account also includes lively descriptions of classical remains and rural society in southern Italy.

Cambridge University Press has long been a pioneer in the reissuing of out-of-print titles from its own backlist, producing digital reprints of books that are still sought after by scholars and students but could not be reprinted economically using traditional technology. The Cambridge Library Collection extends this activity to a wider range of books which are still of importance to researchers and professionals, either for the source material they contain, or as landmarks in the history of their academic discipline.

Drawing from the world-renowned collections in the Cambridge University Library, and guided by the advice of experts in each subject area, Cambridge University Press is using state-of-the-art scanning machines in its own Printing House to capture the content of each book selected for inclusion. The files are processed to give a consistently clear, crisp image, and the books finished to the high quality standard for which the Press is recognised around the world. The latest print-on-demand technology ensures that the books will remain available indefinitely, and that orders for single or multiple copies can quickly be supplied.

The Cambridge Library Collection will bring back to life books of enduring scholarly value (including out-of-copyright works originally issued by other publishers) across a wide range of disciplines in the humanities and social sciences and in science and technology.

Narrative of an Excursion to the Lake Amsanctus and to Mount Vultur in Apulia in 1834

CHARLES DAUBENY

CAMBRIDGE
UNIVERSITY PRESS

CAMBRIDGE UNIVERSITY PRESS

Cambridge, New York, Melbourne, Madrid, Cape Town,
Singapore, São Paolo, Delhi, Tokyo, Mexico City

Published in the United States of America by Cambridge University Press, New York

www.cambridge.org
Information on this title: www.cambridge.org/9781108029636

© in this compilation Cambridge University Press 2011

This edition first published 1835
This digitally printed version 2011

ISBN 978-1-108-02963-6 Paperback

NARRATIVE

OF AN

EXCURSION TO THE LAKE AMSANCTUS

AND TO

MOUNT VULTUR IN APULIA

IN 1834.

BY

CHARLES DAUBENY, M. D. F. R. S.

PROFESSOR OF CHEMISTRY AND BOTANY.

OXFORD,

PRINTED BY S. COLLINGWOOD, PRINTER TO THE UNIVERSITY, FOR

THE ASHMOLEAN SOCIETY.

MDCCCXXXV.

NARRATIVE

OF

AN EXCURSION TO THE LAKE AMSANCTUS AND TO MOUNT VULTUR IN APULIA, IN 1834.

READ TO THE ASHMOLEAN SOCIETY, DEC. 4, 1835.

WHEN we consider the rambling propensities of our countrymen, who every year since the peace have peopled at certain seasons the great roads of the continent, creating a new class of inns for their express accommodation, and introducing their own usages and fire-side comforts into the most distant extremities of Europe, it is singular, how few ever deviate from one or two beaten tracks, and what a scanty measure of information is doled out to us, with respect to places, that lie only a short distance apart from those lines, along which the tide of modern travelling is accustomed to flow.

So long as our inquiries, whether they chance to be of a scientific or of a literary description, are limited to a certain established range of country, we are absolutely embarrassed by the multitude of guide-books and travels that recommend themselves to our notice, and find ourselves provided with more minute instructions, as to the objects of curiosity existing in many parts, for instance, of Italy, than we are with respect to those of our own country.

But when we wish to step a little out of the beaten path, and desire to explore those parts of the continent, which, though lying neither in the direct road to Vienna, to Naples, or to Switzerland, are yet not without their own peculiar sources of interest, we often find it difficult to ascertain even the commonest particulars, with respect to the inhabitants, their roads, or their accommodations; and regard the most scanty and jejune account, which we may chance to discover, respecting such places, in the light of an important acquisition.

It is with this feeling, that I now offer to the Society a brief sketch of a tour, which I performed last winter through a part of Apulia, which, although abounding, not merely in curious natural phenomena, but likewise in picturesque beauties, such as at a better season would render it highly attractive to every lover of bold and romantic scenery, has scarcely to my knowledge been described in any of the many books of travels that have issued from the British press since the peace of 1815.

Mr. Fuller Craven, indeed, in his Tour through the southern Provinces of the Kingdom of Naples, has noticed a portion of the road; but he has almost entirely omitted mention of the two points which above all others attracted me to that country, I mean the Lago d' Ansanto and the extinct volcano of Mount Vultur; and sir R. Colt Hoare, who published his Classical Tour with the intention of supplying those blanks which existed in the descriptions given of the interior of Italy, was prevented, by the inclemency of the season, from proceeding further on the road to Apulia than Benevento.

In order therefore to render my Narrative of some use to those, who may be induced, though with different views to mine, to visit, the classical country of the Samnites and the Hirpini, the birthplace of Horace, or the field of Cannæ; all objects that might be readily comprehended in a tour similar to one I accomplished; I shall on the present occasion, not merely state the results of my inquiries into those natural phenomena which lay in my way, but throw the whole into the form of a journal, extracted, with some few additions, from the diary made on the spot [a].

———

On Wednesday, December 3, 1834, having engaged a Neapolitan Vetturino for the whole of my projected journey, I started from Naples, and proceeded by the Porta Capuana along the great road made in the reign of Charles V. through Apulia, which being the only one existing in the whole country, excepting that to Rome, excited great admiration at the time, and is celebrated by elaborate Latin inscriptions placed upon blocks of stone, that still stand on the road's side, along a considerable part of the line, at intervals of only a few miles.

The first portion of the way lay along a level plain, which in the vicinity of Naples is marshy and unwholesome, from the overflowings of the waters of the Sebeto, once a considerable stream, but now, from the changes effected in the level of the country by the eruptions of Vesuvius, contracted to an insignificant rivulet, which loses itself chiefly in swamps, without reaching the sea. As we proceeded, the country on either side appeared to increase in fertility; and though we were in the midst of winter, yet it was easy to imagine the luxuriance which it must present in autumn, when the elms and poplars, which extend along the whole of this great plain to the very foot of the Appennines, continue in

[a] For the distances, see the post road given in the Appendix.

full verdure, and are united by festoons of vines, stretching from tree to tree, and loaded with a profusion of fruit.

The substratum is made up of a volcanic tuff, such as we see every where in Campania, the wide extent of which plainly indicates, that volcanic action had been rife in the country, long before Vesuvius began to emit lava, and when the country was still under water.

Near the village of Cisterna, a few miles from Naples, I perceived a quarry of leucitic lava, very similar to that of the Monte Somma, and evidently constituting a stream which had descended from that mountain, and not from Vesuvius. Since indeed the focus of eruption has been transferred to the present crater, the whole of the northern skirts of the mountain have been screened from lava-currents by the intervention of the M. Somma, the summit of which constitutes a great semicircular wall, having betwixt it and the crater of Vesuvius, the valley known by the name of the Atrio del Cavallo. The existence of this lava-current would therefore in itself furnish a sufficient proof, were any wanting, that antecedently to the famous eruption in the reign of Titus, to which we are led to attribute the commencement of the eruptions of Vesuvius itself, and the breaking away of the great crater, of which Monte Somma still forms a part, lava-currents had been given off from that part of the volcano, which is now in a quiescent state.

Having nearly reached the foot of the Appennines, we turned off from the direct road to visit Nola, a town of much classical repute, but at present presenting scarcely any remnants of antiquity. Those who, like myself, are induced to diverge from their route in order to visit it, will do well to inspect a collection of Etruscan vases, belonging to a priest resident in the place, which, though not large, appears extremely choice.

After leaving Nola, we soon began to ascend the Appennines, a mass of limestone for the most part destitute of petrifactions, though generally regarded as coeval with that of the Jura mountains; which latter again appear to have been identified in point of age with the oolitic rocks of England.

This range of mountains stretches like a great wall from the north-west to the south-east, extending from the Alps to the very foot of Italy, but leaving on either side of it a considerable tract of comparatively level ground, of which the Campagna di Roma, the Pontine marshes, and the Terra di Lavoro round Naples on the west, and the flats of Taranto, Bari, and Otranto on the east, may be cited as examples.

That the tuff which we found on the plain should also extend to a considerable distance up the slopes of the Appennines, I was fully prepared to expect from having observed the height which the same stratum attains near Naples, at the Camalduli, and in other places.

But it is interesting to be able to trace proofs, in the relative position of the Appennine limestone and the volcanic tuff, of the elevation of the former rock, and even its excavation into valleys, having taken place, prior to the formation of the volcanic matter which rests upon its flanks. This I infer, from the appearance of the limestone jutting out, and, as it were, protruding itself in several places, at an elevation much inferior to that which the tuff is seen to attain; and likewise from the occurrence, high up on the chain, of lateral valleys, the summits of which are capped with limestone, though the hollows and sides present only tuff. The sides of the road which we ascended afforded us an excellent section of the materials of which this volcanic deposit is composed. I found it to consist of an alternation of beds, of volcanic tuff, and of pumice of various colours and consistence. The great thickness which the mass must have attained, when it extended over the plain of Campania, as it once appears to have done, to a considerable height above its actual level, and the vast number of strata which might be counted on our ascent, are circumstances which forcibly arrest the attention of the scientific traveller, but which more properly belong to the general history of volcanos, than to the present detail of local phenomena [a]. I may however remark, that this filling up of valleys previously existing in the older rocks of the country, by volcanic tuff, is by no means peculiar to this locality; it is observable at Sorrento on the one side of Naples, and at Caserta on the other; but nowhere perhaps is it so well seen as on the road into Apulia, both from the many miles over which it there extends, and from the section constantly exposed by the great military road which intersects it.

Having reached the summit of the Appennines, we were rewarded by a magnificent view on either side: on that of Naples, the Terra di Lavoro, unrivalled for fertility, lay stretched before us in all its beauty; but still more striking was the prospect on the side of the Appennines.

On a hill to the right is the picturesque ruin of an

[a] See Appendix.

ancient fortress, called Monte Forte, very formidable in the middle ages, and also celebrated, if I was rightly informed, as the spot where the last ill-fated Neapolitan revolution first broke out.

On a much loftier hill to the left, called Monte Vergine, the sides of which are clothed with chestnut trees, formerly was the site of a temple of Cybele; now, without any diminution of its former sanctity, dedicated to the Madonna, to whose honour a monastery has been erected on the summit.

Once a year, thousands from all parts, regardless of the steepness of the ascent, toil up the mountain, either to offer their adorations at the shrine, or under the influence of the various human motives which conspire to swell the numbers collected on such occasions.

Lastly, in front of us was stretched the elevated table land of the province of Principato Ultra, which, although diversified every where by valleys and acclivities, is throughout raised considerably above the level of the great plain of Campania.

Accordingly our descent was much less considerable than our previous ascent had been; and after reaching the little village which lies under the old fortress of Monte Forte, an avenue of poplars, nearly a mile in length, brought us to the town of Avellino.

Avellino itself is a large and populous place, the capital of the province, and the residence of the governor or intendant. It is now, as of old, famous for its orchards and its hazel-nuts. Virgil celebrates the former,

Et quos maliferæ despectant mœnia Abellæ;

and the name of nux Abellina given to the hazel-nut indicates the abundance of that tree in its neighbourhood. To the stranger, however, it presents few objects of interest; and all I need remark further respecting it is, that the traveller will meet there with a very tolerable inn, and that he will find it convenient to make it his starting-place for the Lake Amsanctus, which may be reached from it on horseback in little more than half a day, thus allowing time, in the fine season, to proceed onwards before night-fall to Ariano, where decent accommodations may likewise be secured.

Being encumbered with a carriage, I was compelled to take a more circuitous route, which traversed an upland, and, for the first part of the way, an highly picturesque country, passing through Pratola and Mirabella, where I diverged from the main road, in order to reach the little town of Frigento, situated about five miles to the east of it.

For some miles after leaving Avellino, I was surprised at observing a continuation of the same strata of volcanic tuff which I had traced all the way during my ascent; so that, as it is quite evident that no volcano exists in the neighbourhood, from which they could be derived, we can only suppose that the waters which washed this material into the Appennine valleys, as before noticed, extended to such a height as to spread it over the crest of the mountain chain which we crossed on reaching Avellino. All traces of volcanic matter appeared however to have ceased several miles before we turned off from the main road at Mirabella, and nothing appeared but the blue compact limestone of the Appennines, alternating with beds of grit, but uncovered by any other material.

Frigento itself, like most of the towns in Apulia,

is built on the brow of a hill, and is inaccessible for carriages: leaving ours, however, at an osteria below, we ascended it on foot, and found there an hospitable reception in the house of Don Martino, one of the principal proprietors of the place, who had been prepared for our visit by a letter from a relation at Naples, and was looking out for our arrival.

Nothing, I am sure, could exceed the hospitality and kindness we experienced during our stay from this gentleman and his brother the Canonico; but for the sake of those who may follow a similar route, I will remark, that in this bleak and exposed country, December is not the month to enjoy comfort in the country houses of the Neapolitan gentry. In all parts of the world the habits and fashions of the capital are more or less a law to the provinces; and as the Londoner has introduced at Naples the grates and coal fires which he uses at home, so in the little town of Frigento, the very name of which (Frigento, a frigore) indicates its bleakness, the casino of my friend Don Martino was built after the fashion of those in the plain of Campania, and boasted not a single fire-place in any one apartment excepting the kitchen.

There accordingly the whole family (for they were by no means Stoics as to cold) crowded after their evening repast was concluded, and being joined by a number of their neighbours, who had been attracted by the uncommon sight of an English traveller, formed a considerable circle round the hearth.

When we retired from this room, a brazier was the only mode of obtaining warmth that remained for us; and it is one which, except in such severe weather, we should hardly have wished to resort to, from the oppressive and unwholesome fumes which attend its use.

The next morning I proceeded to the lake or pool of Amsanctus, which is situated in the plain below, about midway between Frigento and another similarly situated town called Grotta Maina [b].

The lake and valley of Amsanctus is familiar to every schoolboy, as the spot through which Virgil represents the fury Alecto, after having excited discord between Turnus and Eneas, as descending to the infernal regions.

Some indeed have suggested, that Virgil alluded to certain sulphureous ponds that occur near Rieti, above the famous falls of Terni, where the immense accumulation of travertine proves, that the extrication of carbonic acid, and probably of other gases usually accompanying it, has been taking place on an enormous scale ; but the locality seems to be fixed by a passage in Cicero, who expressly says, that the Lake Amsanctus lies in the country of the Hirpini, "Quid enim? non videmus quam sint varia terrarum genera? Ex quibus, et mortifera quædam pars est, ut et Amsancti in Hirpinis, et in Asia Plutonia." A passage in Pliny might also be quoted to the same purpose, but the description which Virgil gives is so applicable to the spot, that I conceive no doubt can remain as to its identification. Thus Ansanto stands "Italiæ medio," (midway between the Adriatic and Mediterranean,) "sub montibus altis," (having in its immediate neighbourhood the hills of Frigento, Villa Maina, &c., and at a still greater distance, the mountainous heights of Monte

[b] Mr. Hamilton has published in the third volume of the Geographical Journal an account of his visit to this spot.

Vergine, and other of the loftiest points in the chain.) It stands in a little valley, which till lately was flanked on either side by a thick wood,

> densis hunc frondibus atrum,
> Urget utrinque latus nemoris,

for it was only in the time of Murat that the wood on the right hand was cut down; whilst a small mountain torrent flows down the valley, and supplies the lake,

> medioque fragosus
> Dat sonitum saxis, et torto vortice torrens.

Here from the pool, which occupies a particular spot where the valley is somewhat wider than elsewhere, issues an enormous quantity of noxious gas, whilst the same proceeds copiously from a cavity in an adjacent rock, so that the latter part of the poet's description is exactly verified,

> Hic specus horrendum, sævi spiracula Ditis,
> Monstratur; ruptoque ingens Acheronte vorago
> Pestiferas aperit fauces.

Lastly, they point out on an adjoining eminence some vestiges of the neighbouring temple to the goddess Mephitis; and Romanelli notices an inscription found on the road from Ariano to Montecalvo, which makes mention of a votive offering presented to the goddess by a Roman lady,

> Paccia Quintilla Mefiti votum solvit.

In our visit to this place, after we had descended the hill on which Frigento stands, we shortly came upon a small pool of water which, like the larger one to which attention has chiefly been directed, emits a quantity of gas. This I collected in the usual manner, by inverting a jar full of water over the spot, on the surface of the pool, from which the bubbles issued; and on my return to Frigento in the evening, submitted it to chemical examination, and found it to consist partly of carbonic acid, and partly of sulphuretted hydrogen. After the removal of these two gases, however, there was a small residuary quantity of air, which contained about 16 per cent. of oxygen and 84 of nitrogen.

We next reached the narrow valley, or rather watercourse, in which the larger pool of Amsanctus is situated. Nearly two miles below it is a warm spring, which disengages carbonic acid and sulphuretted hydrogen gases, and which in the lapse of ages has produced on either side of the water a thick deposit of travertine.

A little beyond us, towards the south, was the town of Villa Maina, standing on an eminence, which I am induced to notice from a circumstance, stated to me by a resident gentleman, who was of the party, respecting it, which seems worthy of notice.

He assured me, that the health of the inhabitants had appeared to suffer materially, since the cutting down of the wood which lay betwixt them and the mephitic lake; that they were sensible of the odour proceeding from it when the wind blew from that quarter, which was not the case before; and that they attributed the insalubrity of the situation to that circumstance. My informant further assured me, that the inhabitants of Villa Maina are noted for the sallowness of their complexions, and that the disease to which they are chiefly subject is an affection of the liver.

Fanciful as it may appear to connect any disorder of the animal functions with the minute quantity of a deleterious gas, which can be so wafted to a distance, so great as four miles from the place of its emission, I am not disposed altogether to pronounce it impossible; for without being a convert to the doctrine of homœopathy in the extent to which it is carried, I can conceive, that the long continued action of a very minute quantity of a noxious ingredient may sensibly influence the system, whilst an infinitely larger dose, applied during a short space of time, may be thrown off without inconvenience, in consequence of the resistance which the vital principle opposes to its introduction, or, as our predecessors in the medical art preferred to denominate it, the *vis medicatrix naturæ*. Thus it is at least not absurd to suppose, that an individual who had exposed himself for some hours to these vapours within the crater of Vesuvius, in the most concentrated state in which they can be endured without subsequent injury, might be affected by inhaling, during many years, a very minute portion of the same, wafted from the Lake Amsanctus, whenever the winds set in that direction.

The quantity of mephitic vapour which proceeded from that quarter was such, as to oblige us (the wind being in the north) to take a circuit towards the east, in order not to meet the noxious blast; instances not unfrequently occurring of animals, and even men, who have imprudently ascended the ravine, being suffocated by a sudden gust of air wafted from the lake.

This is the origin of the fable of the Vado Mortale, a particular spot in the course of the rivulet that flows from the lake, which, it is said, cannot be forded without death, and which has been described, as having on its borders an accumulation of the whitened bones of the various animals that had

stone, a bed of gypsum nowise different from those occurring in many tertiary deposits.

Now it was interesting to me to remark the analogy between the appearances presented here, and those at the Lagoni near Volterra in Tuscany, which I had visited on my way to Naples just before. There, too, volcanic rocks are entirely absent; and the formation for many miles round the spot consists of the Appennine limestone, with serpentine or diallage rock protruding through it.

But at the Lagoni we see pools of water in a state of absolute ebullition, from the quantity of steam which is constantly rising through them, and which imparts to them a temperature exceeding 180° of Fahrenheit. This steam seems to carry up with it boracic acid and sal ammoniac; the former in sufficient quantities to make it worth while to evaporate the water through which it passes in order to collect it; the liquid being conducted into shallow troughs, where it is mixed with soda, by which addition crystals of borax are obtained from it as the aqueous portion escapes.

Now when we compare together the effects produced by the disengagement of steam and sulphuretted hydrogen, owing almost unquestionably to a volcanic cause, in the instance before us and in that of the Lake Amsanctus, we are naturally led to apply the same explanation to those immense deposits of sulphur which occur on the western side of Sicily. If any doubt should exist as to the fact of their having been so produced, it may be removed by reflecting, that we know of no instance of sulphur being sublimed in an uncombined form by volcanic action, and that it seems scarcely possible for such an event to occur, without the combustion of the sulphur taking place the instant of its coming into contact with atmospheric air.

Hence it may be inferred, that the whole of this vast deposit in Sicily has been occasioned by a decomposition of sulphuretted hydrogen gas, such as has taken place on a smaller scale at Lake Amsanctus, and has there impregnated the surrounding rocks with crystals of the same material.

In Sicily too we meet with all the combinations which sulphuric acid is capable of forming with the earths present—in the beds of sulphate of lime, of strontian, of barytes, of magnesia, that occur—there also we see in the immediate neighbourhood warm springs impregnated with sulphuretted hydrogen— memorials, as it were, of the cause, which had produced these deposits.

May we not also be led to conjecture, that the gypsum so commonly present in the tertiary clay of Volterra and the maremnæ of Tuscany, has been produced by the same process, especially when we find this clay associated, as it frequently is, with beds or nests of sulphur.

Thus the formation of the blue clay, in Sicily and in the maremnæ of Tuscany, would have taken place, not only at the same epoch, which is generally admitted to have been the case, but under the same physical conditions, one as the other, and a volcanic action similar to that going on at the Lago d' Ansanto and at Monte Cerboli would likewise have given rise to the deposits, which the former contain, in common with the rocks found immediately round the spots, where the above operations are at present proceeding.

But there is another circumstance also worth noticing, although the inference to which it seems to point will scarcely receive admission until further proof can be adduced in support of it. I allude to the association of salt springs with gypsum and sulphur both in Tuscany and Sicily. Their occurrence in such localities as these might induce us to conjecture, that the same volcanic action, which produced the sulphuric salts, and volatilized the sulphur, has been instrumental also in separating the salt from its solution in water, and thus serve to explain, in these instances at least, the puzzling fact, that rock-salt is found associated, as is so commonly the case, with beds of gypsum. The connexion between the above phenomena may perhaps be seen more clearly by the following tabular view:

Volcanos give out...... ...... { sulphuretted hydrogen, sal ammoniac, boracic acid, muriatic acid, steam;

and cause .... { deposits of sulphur, of sulphuric salts, of muriatic salts, &c.

Moffettes, connected geographically with volcanos either now in action, or extinct, give out.............. } the same principles,

and cause..................... { deposits of sulphur and of sulphuric salts.

Many tertiary clays, some of which are connected geographically with volcanos, contain..................... } beds of sulphur, of earthy sulphates, and of common salt.

Most salt formations are associated with................ { beds of gypsum,

some with....................... sulphur,

others with .... sal ammoniac.

To return from this digression to the Lake Amsanctus.

It is interesting to remark, that the position of this spot is almost exactly intermediate between the active volcano of Vesuvius and the extinct one of Mount Vultur, and that a straight line drawn from the one to the other, and which might likewise be extended on to the volcanic island of Ischia, would pass within a mile or two of this mephitic lake.

From the neighbouring eminence on which Frigento is situated, the summits of both mountains were alike visible; and it was stated to me on the spot, that when Vesuvius was in a state of activity, an unusual quantity of gas was disengaged from Amsanctus. Hence it seems probable, that the same elements of volcanic activity exist underneath the earth all across the peninsula, although these elements may be called into more intense action on either side of the Appennine chain, by the proximity of sea water, or by some other circumstances which do not occur in the central portion of the chain.

The occurrence of Plutonia, or places which disengage large volumes of noxious gas, is familiar in all volcanic countries, but I know of no other instance, in which the quantity is considerable enough to endanger life in the open air round about the point of its emission, except that of the Valley of Death in Java [a].

---

[a] It is a valley near the summit of a mountain, about half a mile in circumference, the depth from 30 to 35 feet, the bottom quite flat, no vegetation, a few large (in appearance) river stones, and the whole covered with the skeletons of human beings, tigers, pigs, deer, peacocks, and a great variety of birds and beasts. The party who visited it could not perceive any vapour or

together, and to have converted the ridge of the
Appennines into a succession of islands. In fact
the principal passes which traverse this chain of
mountains, are carried over eminences of considerable height, such as that of Abetoni between Pis-
toya and Modena, which has an elevation, according
to Pini, of 4166 feet; that of Pietramala between
Florence and Bologna, Furlo near Fossombrone, Colle
Fiorito between Terni and Foligno, Bocchetta in
Liguria. In order to determine with precision the
length of this ancient peninsula, and to mark the
spots where its continuity was broken, we ought to
ascertain what was in different places the elevation
and depression of the central portion of the Appennines, previously to the corrosions and rents which
the rivers and torrents had in after-times effected.
One of these interruptions certainly existed in that
part of the kingdom of Naples, which bears the
name of the Principato Ultra, and was in a valley
which is prolonged from Avellino to Bovino, and
which leads from the Terra di Lavoro into Apulia.
This valley at one time cut through those Appennines to which belong the gigantic calcareous mountains of Incoronata, Monte Vergine, and Avella, for
if on the side of Ariano, or in its central portion,
it is seen to be closed in by great eminences, the
latter have been formed by deposits analogous to
those of the hills of Tuscany and the Romagna,
consisting in a calcareous sand with grains of silex
and scales of mica, which at Monte Reale, between
Bovino and Ariano, as likewise between this latter
place and Grotta Minarda, is seen to be superimposed
on a marly bluish clay. This sand extends from
the side of Apulia nearly to Ordona, and is pro-

longed beyond Naples, though near Pratola it begins to be covered with a volcanic tuff derived from the substances erupted from the submarine volcanos of Campania, or transported by the waters which insinuated themselves into the hollows and recesses of the surrounding mountains.

"'The valley of Bovino then separated from the rest that tract of country comprehended under the Basilicata, Apulia, and Calabria, which then appeared like a distinct island, or rather was an union of islands formed by the highest peaks of these mountains, since the Murgiæ of Apulia, and the low eminences, which, branching through the territory of Otranto, extend as far as Cape Leuca, then continued submerged."

On this description of Brocchi's I would only observe, that the whole of the valley spoken of must not be inferred to be covered with tertiary deposits: on the contrary, the line of road I pursued from Avellino to Frigento, and from Frigento to Ariano, presented only the Appennine limestone with its subordinate beds, until we came within a few miles of the latter town, which stands on a lofty eminence, composed entirely of tertiary beds, resting on a loose conglomerate.

The softness of the stone of which the rock consists has rendered it so easy to hollow out, that the whole surface towards the south is honeycombed into caves, which form the dwelling-places of a large proportion of the population, as is the case in one or two places in the tertiary deposits of central France, at Lisle, &c. It is remarkable that the people, and more especially the children, who emerged from these damp and sombre habitations, seemed

perfectly healthy, contrary to what has been observed in some other places of the kind, where, in
consequence, as is supposed, of the absence of light,
the form and functions of the animal system are but
imperfectly developed [c].

From Ariano we proceeded along the valley till
about nightfall, when we reached the bridge of Bovino, the town itself standing on an eminence
above.

This spot, I may remark, was formerly infamous
for the robberies committed in it; the locality being
peculiarly favourable for such enterprises. For the
last few miles the road lay in a narrow glen, hem-
med in on either side by a steep ridge, covered on
the top with wood, from whence the robber could
watch his prey, and plan his attack at leisure. Add
to which, that there were no means of escape, except
through the pass at which the bridge of Bovino
stands, and that for several miles before reaching
it, there is no track to the right or left branching
out of the main road, by which the traveller might
avoid the danger awaiting him.

Even now, that the bands which formerly infested the country are destroyed, videttes continue
to be planted at distances of a mile or two along the
whole of the pass, and few choose to delay their
journey through it till the close of day.

Finding only a most filthy inn at the bridge of
Bovino, and not relishing the idea of being benight-
ed in such a country, we were fain to put up at an
osteria a few miles beyond, where we should have

[c] See Dr. Edwards' experiments on tadpoles, which, when
secluded from light, never underwent their metamorphosis into
frogs, though they continued to increase in size.

fared but badly, had we not brought with us our own provisions, and been able to avail ourselves of our carriage as a retreat for the night.

At the bridge of Bovino we took our leave of the Appennine chain, and found that we had begun to enter upon the wide plain, which extends to the shores of the Adriatic, a distance of nearly thirty miles. This tract seems to be wholly composed of very recent deposits, and to have constituted at a comparatively modern epoch, a gulf of the sea, which was at that period circumscribed, on the south-west by the range of hills which stretches from Bovino to the neighbourhood of Melfi, and is from thence prolonged to the present gulf of Taranto, and on the north by the heights of Mount Gargarus in Manfredonia ; both equally belonging to the Appennine chain, and consisting of older formations.

At that point, we quitted the great road into Apulia, which had brought us thus far with ease and safety, and had to plough our way as best we could, through a wild uninclosed tract of land, chiefly devoted to pasturage, but, so far as relates to roads, and other evidences of human cultivation, in a condition little more advanced than when it first emerged from the ocean.

To those who are disposed to follow my footsteps, I may suggest, that this part of the journey would be accomplished far more expeditiously, as well as more agreeably, on horseback, in which case the traveller might, if well-mounted, reach Melfi, the nearest town to Mount Vultur, in one day from Bovino, by setting out at an early hour.

Encumbered by a carriage, and travelling at a season, when the sandy loam over which we had to pass, was rendered stiff and heavy by the rains that had fallen, we could get no further that day than Ascoli, a wretched town, where we procured very sorry lodgings at a little inn kept by an old woman, possessing a strongly marked physiognomy, but whose haggard looks and wild expression of countenance might have suited one of the witches in Macbeth. Being curious to learn her history, I made inquiries of a person of the town with whom I chanced to converse, and was told, that she was the sister of one of the famous robber-chiefs of Apulia, and in her youth had possessed considerable personal charms; that she had lived with her brother and his troop during many years in their mountain fastnesses, and had shared their prosperous days; but when the band was broken up, and her brother had been apprehended and executed, had found it necessary to take up with the occupation of an innkeeper, which in my informant's opinion was a great sinking of caste. His feeling on this last point will seem more intelligible, when we recollect, that at a time not very remote, the leaders of the banditti in Apulia were generally persons of some rank; that they often combined a wild feeling of chivalrous honour with their lawless habits of life; and that whilst their power made them objects of terror to all, their system of confining their outrages to the rich, created a kind of fellow-feeling in their behalf amongst the humbler classes.

Ascoli, the ancient Asculum, is celebrated for the second battle between Pyrrhus and the Romans, which, though hotly contested, terminated without any signal advantage on either side. It stands on a hill, based on a shining, glary, sandy clay, upon which rest beds of yellow sand; the whole being capped with thick beds of conglomerate, in which rolled pebbles of the Appennine limestone, and of some other rocks, none of which however are volcanic, are firmly agglutinated together by sandstone, so as in some places to constitute a very coherent mass. I could perceive no organic remains in any part, but have no doubt, from its general aspect, that it is of tertiary origin.

The next morning we prosecuted our route to Melfi, passing at first over a plain as wild and as uncultivated as that we had traversed on the day preceding. About nine miles from Ascoli we crossed the Ofanto, the Aufidus of the ancients, a river at the time we passed it of very inconsiderable size. We had next to ascend several pretty lofty ranges of hills, from the summit of the last of which the town of Melfi was visible.

From this point the view that opened upon us was singular and beautiful; at the extremity of the landscape stood the Mount Vultur, which in its general outline bears a striking resemblance to Vesuvius, as we imagine it must have been before it broke out in A. D. 79.

Between the spot where we were and the mountain, stood the town of Melfi, built upon an isolated neck of land, with its castle, a picturesque building, overhanging the precipice. A little stream winds round the valley, dividing us from the rock on which Melfi is situated, and a bridge thrown over the stream at the bottom of the valley connects the town with the road. No volcanic appearances presented themselves until we reached the range of hills nearest to the former, previously to which, the tertiary sand and clay of the neighbourhood of Ascoli, covered with puddingstone, containing pebbles of the Appennine limestone, alone made its appearance.

But the range immediately south of Melfi seemed to be based upon the older Appennine limestone, covered over however by beds of tertiary sandstone capped by volcanic tuff, so that, as we descended the valley intervening between this eminence and the town of Melfi, we soon found ourselves travelling upon that limestone.

The hill, however, upon which the town itself is built, consists of different materials. It is based upon beds of volcanic tuff, which (as may be seen in the annexed drawing) dip from the Mount Vultur, so that at the north western side of the rock of Melfi they lose themselves under the surface. These are covered by a thick bed of lava, which presents a precipitous escarpment, as well as an imperfect columnar structure, towards the north-west, and appears to have proceeded from Mount Vultur, though the changes which the country has since undergone may have severed the connexion with its parent.

At Melfi I found tolerably comfortable accommodations at the Nobile Locanda di Sole, and every possible attention from the gentlemen resident there to whom I had brought letters.

The town appears not to have existed in the times of the Romans, but to have been a place of some repute in the middle ages, when the Normans, under Count Ranulph, seized and fortified it. The investiture of the Dukes of Apulia accordingly took place there, and it was at that time considered one of the principal cities of the province.

It is at present a large but an extremely dirty town, principally belonging to the prince of Doria.

It lies so much out of the general route, that the intercourse with foreigners is very small; and the people, though honest and obliging, are ignorant and superstitious to a degree which the stranger will do well to be aware of.

Almost the last scientific traveller before myself, who made Mount Vultur the object of his researches, was the celebrated Italian geologist, Brocchi, who, burdened, not only with a hammer, but likewise with a barometer and other philosophical instruments, proceeded some fifteen years ago to ascend the mountain.

The peasants, at a loss to conjecture the nature of his objects and his manipulations, set him down as a magician, but not knowing whether he might have come to do them good or harm, contented themselves at first with watching him, very attentively, but at a respectful distance. Unfortunately, in the very midst of his observations, the heavens lowered, the wind betokened an impending tempest, and drops of rain began to descend.

The peasants regarded all this as the first-fruits of his incantations, and awaited with silent dismay the result; but when they saw a furious thunder-storm invading their crops, and demolishing their hopes of a coming vintage; whilst the philosopher continued inspecting with redoubled attention those instruments of his craft, which had been regarded by them before with so much suspicion, fear gave way to indignation, and they rushed forwards in a body with the full intent of demolishing the mysterious author of all this mischief.

Fortunately for Brocchi, he had with him three or four resolute gens d'armes, as his escort; and these, with their muskets and bayonets, contrived to keep the unarmed peasants at bay long enough to enable him to escape, or his zeal in exploring the secrets of volcanic action might have been as fatal to him, as it was of old to Empedocles[f].

The following morning I ascended mount Vultur in company with a gentleman of Melfi, who insisted on being of the party, and two gens d'armes who were sent as an escort, in addition to the ordinary guides, &c. We first wound round the western flank of the mountain, keeping on our right hand the valley, at the bottom of which flows the river Ofanto. Wherever the substratum was exposed, it seemed to consist of volcanic tuff, of which various beds were visible, some very compact, others loose and friable, consisting chiefly of pumice like those about Pompeii [g].

In this loose material we saw several caverns excavated by art, some of which, we were told, had been the lurking-places of those bands of robbers, who within the last twenty years laid the country

[f] According to a recent tourist in Ireland, a similar adventure lately befell a botanist herborizing in the mountains of Connemara during the time the cholera was raging.

He was nearly murdered by the country people, in consequence of being mistaken for a French doctor sent by government to inoculate the people with this dreaded plague; and the tin box, in which he carried his plants, was regarded by them as the case, in which his stock of infection was kept, in readiness to disseminate wherever he went. See Angler in Ireland, vol. i. p. 189.

[g] The neighbourhood is said to yield a profusion of plants, and being almost unexplored by modern botanists, is likely, if visited at the proper season, to yield a rich harvest to the collector. Like most volcanic soils, it produces excellent wine.

under contribution, and made mount Vultur their principal retreat.

After skirting the base of the mountain in this manner, we came at length to the wood, on its southern flank, termed Montecchio, and, having crossed a deep ravine, in which flows a ferruginous water, tinging the stones over which it passes with a red ochreous sediment, entered an extensive forest, well calculated to be the haunt of the numerous banditti who formerly resorted there.

In the midst of this forest is a spring called the Aqua Santa, quite cold, and without any particular taste, but possessing some reputation in the cure of diverse disorders, and giving out bubbles of air. This I collected, and ascertained to consist of carbonic acid gas, together with a small portion of residuary air, amounting to about 1 p^r. c^t., 100 parts of which contained 10 of oxygen, and 90 of nitrogen [h]. The same gas escapes in still larger quantities from a fissure in the rock hard by, and a considerable deposit of travertine in the neighbourhood indicates, that it has been given off elsewhere [i].

From this point we began to ascend in a northerly direction, and upon reaching a particular point about half way up the mountain, found ourselves in the midst of an amphitheatre of hills, which from its form, as well as the physical constitution of the

[h] Thus confirming the inference which I have deduced from many former observations, that whenever atmospheric air escapes from a volcanic source, it is found deprived of a portion of its oxygen.

[i] The same gas is emitted at a place called Rendina, to the east of mount Vultur, and likewise, together with sulphuretted hydrogen, from some mineral waters near Atella, on the western side of the mountain.

rocks surrounding it, immediately suggested to us the idea of its having been at one time the crater of the volcano. The cavity was nearly circular, and the brim, with the exception of the side by which we had ascended, was almost entire.

Its interior was covered with a rich black soil, having but few blocks of volcanic matter scattered over its surface. The rocks encircling it are very irregular in point of height, some of them rising more than 1000 feet above the average level of the margin. One of these eminences, which partakes of a conical form, constitutes the summit of the mountain; for, as is the case in some other instances, there is no crater on the highest point.

Within this great circular expansion, are two minor depressions, which are likewise nearly round, and, as they occupy the lowest portions of the crater, receive the drainings from other parts. Hence they form two lakes communicating by a narrow outlet one with the other, and discharge their superfluous waters by a little rivulet which runs from the lower or more southern. The water of the latter is perfectly sweet, that of the most northern one is stated to be salt, but it must be slightly so, since fish abound in it.

It is said, that at times volumes of sulphuretted hydrogen, or some other inflammable gas, are given off from this lower lake; and jets of water have been known to be suddenly thrown to the height of fifteen feet above its surface. These circumstances evince, that the volcanic action is not completely spent, and still shews itself at intervals, in an energetic manner, by the evolution of gas. On the borders of the upper lake are the ruins of a church, dedicated to S. Hippolito, and on an eminence overhanging it is a convent of Franciscans, who, protected by the sacredness of their calling, and their supposed poverty, continued to reside without molestation in this sequestered spot, at a time when the mountain itself was the head quarters of the bands of robbers which infested the province. Once a year Melfi pours forth its whole population on a pilgrimage to this sacred spot—prompted by the double motive of paying their adorations at the shrine, and of commemorating the escape which a portion of the inhabitants experienced in 1528, by secreting themselves within the crater, and thus eluding the fury of the French troops under Lautrec, who took the town by assault, and put all that remained in it to the sword [k].

The rocks, which encircle this crater, seem generally to consist of tuff, but the substratum is so completely covered with vegetable soil, that it is difficult to determine its stratification. One side of it is covered with a thick forest inhabited by wild boar, the rest is cultivated, and yields abundant crops.

A heavy shower of sleet obliged me, to abandon my original design of ascending to the summit of the mountain, and to hasten homewards.

We accordingly climbed the western edge of the crater at its lowest point, and having reached the summit, descended the opposite side in the direction of Melfi. Not a single stream of lava was observable on the flanks of the mountain on this side; if any such there be, they were concealed by a thick bed of black unctuous soil, resulting from the mixture

[k] Such was the account given me on the spot, but I confess I do not find it confirmed by the narrative of the taking of Melfi furnished by Guicciardini, or by Sismondi.

of volcanic ashes with the clay which proceeded
from the decomposition of harder materials. The
extent to which this has taken place,—the rare occurrence, in a mountain which once was the theatre
of such extensive volcanic operations, of loose blocks,
—and the entire disappearance of streams of lava,
would alone imply the great interval of time that
had elapsed since their occurrence; but another
proof of the same was afforded in the existence of a
deep and wide valley, which we crossed on our way,
extending from the bottom of the mountain nearly
up to the crater, completely covered by vegetation,
and with its superficial strata reduced altogether to
the condition of a rich and slippery loam.

In the few specimens of volcanic rock met with
on our road, either in detached blocks, or composing
beds of loose materials, which the soil had not completely concealed from our view, there was a general
uniformity of character.

They consisted of a dark, and generally compact,
though sometimes cellular, stone, which was made up
of an intimate union of minute crystals of augite imbedded in a felspathic base. There were also dis-
persed through the matrix crystals of a dark green
colour, which, from a comparison with others subsequently found, I conclude to be hauyne.

These masses may either have been detached
from a stream of lava which formerly had been
thrown out, or may have been ejected separately.

If the former were the case, what an idea of antiquity does the complete concealment by vegetation
of these streams convey to the mind, when we recollect the ages required to cover over even with lichen
those of Etna or Vesuvius!

Yet the whole of these eruptions were posterior to the formation of the tuff constituting the base of the mountain, which is of various degrees of compactness, but for the most part resembles that of Posilippo, containing comminuted masses of pumice imbedded, alternating with beds of loose pumice, without any cementing material.

The rock on which Melfi stands consists of the same substances, but hauyne is more abundant.

The best specimens, however, of that mineral occur in a bed of loose volcanic stones, which lie underneath the lava above noticed. When the crystals are capable of being detached from the matrix, they are generally found to possess the dodecahedral form, which in other localities is seen so imperfectly, well defined, and they are characterized moreover by gelatinizing with nitric acid, and by fusing into a blue glass with borax.

The blue kind found at Andernach, and in the Campagna di Roma, is less common here than the dark green variety, and owing to some external decomposition, the former is white superficially, and presents its natural colour only when fractured. It is accompanied with leucite, with augite, and with hornblende, and in the compact lava-bed which overlies the congeries of ejected masses described, I perceived concretions very much resembling the pearlstone of Hungary.

I am unable to discover, that mount Vultur has ever been measured barometrically, but that its height is very considerable there can be no doubt. At Frigento, which is nearly equi-distant from it and Vesuvius, the summit of Vultur appeared more conspicuously than that of Somma over the Appennine ridge. Dr. Bouè, in his recent work entitled "Guide du Géologue Voyageur," assigns to it an elevation of only 1400 feet, but this is manifestly much below the truth. Its circumference is stated at about twenty-two miles.

It had been my original intention to return to Naples by another road, by Muro, Laviano, Oliveto, and Campagna, which traverses some of the loftier peaks of the Appennines, and would have brought us into that from Naples to Pæstum at Eboli.

But we were dissuaded from this undertaking, by the state of the weather, and the probable depth of the snow on the heights which we should have been compelled to pass.

On the other hand, the river Ofanto was so swelled by the rain which had fallen, as to justify the epithet of *violens Aufidus*, which Horace had applied to it:

Dicar, qua violens obstrepit Aufidus,
Qui regna Dauni præfluit Appuli,
  Cum sævit, horrendamque cultis
    Diluviem meditatur agris.

By waiting at Melfi, however, till the flood had in some measure subsided, we were enabled to ford the stream in safety ; and our kind friend, the Cavaliere Araneo of Melfi, whose family had been unremitting in their attentions to me during my stay, enabled my Vetturino to drag the carriage through the mud intervening between the town and the high road, a distance of more than thirty miles, by lending me two good mules, in addition to the three horses which we had brought with us. We therefore returned by exactly the same road which had brought us out, and arrived the fourth day at

Naples, where we found a more genial sky, and a
climate in every way contrasted with that we had
experienced, either upon the Appennines, or round
the skirts of that mountain chain.

I have now only to offer some general remarks on
volcanos, as suggested by the phenomena which
have just been described.

We may, in the first place, point out the existence of a line of country extending from the parallel
of Naples to that of mount Vultur, along which volcanic action is developed under most of its different
phases, according to certain variations in the external circumstances under which it occurs.

In Ischia, where the line commences, hot springs
are abundant, and at distant intervals of time true
volcanic operations have been observed.

At Vesuvius, close to the sea, ejections of stones
and eruptions of lava are taking place continually ;
but when we follow the same line across the Appennines, we look in vain for any distinct indication of
volcanic action, except at the Lago d' Ansanto,
where the same gases are evolved, where a certain
degree of heat seems to be generated, but where
neither stones nor ashes have ever been thrown out.

Having traversed the Appennines, and descended
its eastern slope, we perceive, standing in the same
relation to that chain on the one side, which Vesuvius does on the other, a mountain similar in form
and structure, but evincing no analogous phenomena
at present, except the emission of carbonic acid and
sulphuretted hydrogen, and those occasional rum-
blings of the earth beneath, which have been compared to subterranean thunder, and which, if the

natives are to be believed, take place most commonly when Vesuvius itself is in a state of eruption.

With such evidences of an internal communication, or at least of a sympathy between these two distant spots, what circumstances can be pointed out, to account for the almost entire quiescence of the one, and the determination of the volcanic energy to the other? One remarkable difference existing between them is, that, whilst Vesuvius is close to the sea, Vultur is thirty miles apart from it; yet, as we have seen, this difference may not have existed formerly, since the waters of the Adriatic, if they did not actually wash the sides of mount Vultur, once came within a short distance at least of its base.

And, when we examine the present condition of this volcano, we are led to carry back the date of its activity to a period at least as remote as that alluded to; for, so far from imagining it to have been burning in the times of the Romans, its condition seems to have been at that time but little different from that which at present belongs to it.

Horace would scarcely have selected it as the spot, where the doves are represented as covering him over in the days of his childhood with fresh leaves[1], had not the forests existed then in their present luxuriance; and, if the mountain had appeared to the poet under that arid and forbidding aspect which belongs to all volcanos that have been but a short time extinguished, he would have introduced some

---

[1] Me fabulosæ Vulture in Appulo,
   Altricis extra limen Apuliæ,
      Ludo fatigatumque somno,
         Fronde nova puerum palumbes
   Texere: Hor. Carm. iii. 4.

mention of such a circumstance, when he alludes to the peculiar features of the neighbouring places: the "excelsæ nidum Acherontiæ;" the "saltus Bantinos;" the "arvum pingue humilis Ferenti."

Lucan too speaks of the "arva Vulturis," without any allusion to the harsh and sterile appearance which belongs to a mass of recently ejected lava.

These indications of an extreme antiquity, compared to that of our earliest historical records, will not surprise those who are at all conversant with volcanic phenomena; and perhaps the extinct volcano of Rocca Monfina near Sessa, to the east of the road from Rome to Naples, furnishes an instance at least as remarkable. In the early history of Rome we read, that a war having broken out between the Sidicini and the Aurunci, the latter, through fear, deserted their city, and fortified Suessa, now called Aurunca, the present Sessa, whilst the walls of their former abode, and all that it contained, were destroyed by the Sidicini.

Some remnants, however, of these walls, I am assured, still exist—they are composed of lava, probably derived from one of the streams described by Breislac as having flowed down the mountain, a proof of their priority to the foundation of a city destroyed 334 years before Christ.

But a more decided proof of the antiquity of this volcano may be collected from the circumstance, that the ancient town stood on the brow of a hill, which constituted part of the margin of this very crater, and consequently that the crater which lies beneath it was then, probably in the same condition as at present, namely, covered with vegetation, so as to furnish the inhabitants with a pasturage for their cattle, and certainly not the scene of volcanic phenomena, which would have precluded them from fixing upon such a site for their abode. Thus the period, at which this Volcano was in a state of activity, must be as remote as that, which we are disposed to assign to mount Vultur[m].

Such then are the volcanic phenomena met with in a direct line, but if we take in a somewhat greater breadth of surface, we shall include, on the side of the Mediterranean, the Islands of Ponza, and on that of the Adriatic, those of Tremiti, off the coast of Manfredonia, whilst the earthquakes under which Foggia and other cities have suffered, shew, that the same operations are carried on elsewhere in the neighbourhood.

In the annexed map I have stated the volcanic districts that occur nearly in a line either to the eastward or the westward of the country described, but I do not wish to lay much stress upon this coincidence of position as a proof of connexion between the more distant points, feeling, that there are perhaps few parallels of latitude, which do not present volcanic phenomena in some part of the globe or another.

I am more disposed to view the Italian peninsula,

[m] From a statement of Breislac's which I have quoted in my " Description of Volcanos," p. 137, it would follow that Rocca Monfina had been in action during the historical period, for it is there mentioned that the tuff on which the town of Sessa is built covers the remains of an ancient city, and amongst the rest, the vestiges of an amphitheatre. In my last visit to Italy I examined the spot, and found the buildings alluded to, but the amphitheatre, though considerably below the present town, seems to have been built in a natural hollow in the tufaceous rock, and never to have been covered over by that deposit; neither do the remains of another ancient building that has been detected, bear out Breislac's conclusion, which is moreover encumbered with so many difficulties, as to require the most overpowering evidence to establish its credibility.

as part of a long strip of country stretching from
north-west to south-east, which is everywhere more
or less subject to volcanic operations, but has here
and there particular bands of greater intensity
running at right angles to the former direction, or
nearly from north-east to south-west. The first
might be represented by a longitudinal band, the
second by transverse lines intersecting it. Thus let.
a a a a represent the longitudinal band running
from north to south of Italy, and thence extending
to Sicily, and the transverse line B B will denote
the line of greatest intensity stretching across Tuscany; C C that stretching across the Romagna;
D D that from Naples to mount Vultur; E E that
from the Lipari islands to the theatre of the Cala-
brian earthquakes; F F that from mount Etna to
the new island of Sciacca.

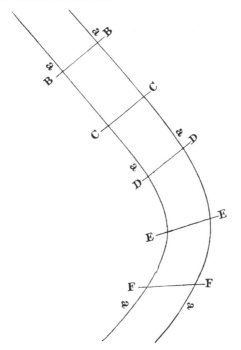

APPENDIX.

Note at page 10.

I CANNOT, however, omit this opportunity of noticing two varieties of volcanic tuff which appear not to have been sufficiently studied, and which shew, that since the original ejection of the materials, their deposition at the bottom of water, and their consolidation into a mass of uniform consistency, changes have taken place in them, which the long continued operation of a moderate heat is perhaps most likely to have produced.

The first of these varieties is seen in the valley of St. Rocca, near the Capo di Monte, above Naples, where the tuff has been extensively quarried, and where large caverns are left in the rock, which, by exposing a considerable extent of surface, allow of our examining the structure on the great scale.

The most remarkable circumstance relating to it, is the existence of veins of a harder variety of the same rock traversing the bed. I call the matter of the veins a variety of tuff, because in some cases there seemed to be a passage from the ordinary kind into it, though in others the line of demarcation is very distinct.

In its hardest form it approaches nearly to porcelain jasper, is splintery, and has a conchoidal fracture. In these cases its surface is smooth, and its texture uniform, but when it is traced on, we discover in it a fragmentary structure, bits of obsidian, porphyry, pumice, and lapilli of various sizes and kinds being imbedded in the same paste. In other instances little veins of the same fragmentary variety of rock pass through the substance of the compact and uniform kind. The harder varieties of tuff, here described as veins, contain specks of mica disseminated through them.

Its surface is sometimes coated with a white incrustation, which seems to be siliceous.

It is seen to rest on a bed of the same volcanic materials, possessing however a very loose consistence, and composed principally of a congeries of grains of sand, and of lapilli.

Of this modification of tuff I know of no notice having been taken, excepting by professor Tenore of Naples, in his "Essai sur la Geographie Physique de Naples;" but the following is alluded to by Breislac and others under the name of Piperno, the name applied to it from one of the places where it has been found.

The spot in which I principally examined this latter variety, was the north side of the crater of Pianura, five miles from Naples. Like several other bason-shaped cavities that occur in the great tufaceous deposit overspreading that country, to which from the similarity of their form we are disposed to attribute the same origin, the sides of this crater exhibit no alternation of beds of lava and of ejected materials, or other traces of having been at any time the theatre of those eruptions of lava which occur at Vesuvius at present. They are composed indeed of the same volcanic tuff as that of Posilippo, with which this bed seems continuous, but the rock, which bounds the crater towards the south, for a certain distance upwards from the level of the hollow within, presents that character which has caused it to be designated by the popular name of Piperno.

Being extensively quarried for building purposes[a], galleries have been carried for a great distance into the rock, both horizontally and perpendicularly, and by means of the latter it has been ascertained that the same material extends at least forty feet beneath the level of the bottom of the crater. With respect to its external characters, Piperno may be described as a sort of volcanic breccia, bearing the same relation to the ordinary tuff, which Sienna marble does to an ordinary puddingstone, containing the same fragments of volcanic matter as the latter, but with the imbedded frag-

[a] It is used for the steps of houses, the sills of windows, &c. being more durable than tuff, and more easily cut than lava.

ments so blended with the cementing material, as to appear like a part of the same.

These fragments, which are sometimes without, but more commonly with cells, are often compressed and lengthened out in an horizontal direction, and are imbedded in a whitish paste, similar to ordinary tuff, but more compact. Indeed the beds of tuff that occur above it resemble much, except in hardness, the basis of the Piperno.

It is interesting to observe, that just over, and, as it were, passing into, this rock, is a tuff, with fragments of these several kinds of lava abundantly scattered through it, but existing in a state of very loose aggregation, excepting in the vicinity of the Piperno, where it becomes more and more compact, gradually passing into a perfect breccia. It is said moreover, that the best and hardest Piperno is taken from the bottom of the bed.

The impression which the appearances of Piperno created in my mind, was that of a bed of ordinary tuff, which after it had been originally deposited, was subjected to the action of heat under pressure, and in this manner underwent a kind of partial fusion. Under this condition of things, crystals of felspar, which are now abundant in it, as likewise the other minerals we occasionally meet with, although they did not previously exist, might be formed, and by the same agency the fragments disseminated through it might become blended with the surrounding matrix in the manner we find them to be.

There is yet another variety of volcanic tuff known in the Campagna di Roma and elsewhere, under the name of Peperino, which must not be confounded with the Piperno already described.

There is, however, thus far a correspondence between them, that Peperino is also a material which appears to have undergone certain intestine changes since it was first deposited, the effects of which have been to produce in the midst of the amorphous mass certain crystalline minerals, such as mica, augite, felspar, &c., and to impart a more uniform degree of compactness to the whole.

Whether these changes resulted from heat subsequently applied, may admit of dispute, but that here also some chemical operation or other has produced a fresh arrangement of the particles since the mass was originally formed, will scarcely be doubted by those, who admit that Peperino is a modification of the volcanic tuff found in its vicinity.

Yet even the latter, as seen at Posilippo, is more compact than the material known under the same name which has covered over Herculaneum, or which has resulted in other instances from the action of rains and torrents upon showers of ashes proceeding from modern eruptions.

Not that there is any original difference between the materials ejected, but that in the former instance the pressure of so much larger a bulk of matter, and probably also of the body of water, which for a time covered it, has caused a more intimate union between the several parts, and given it more of the character of an homogeneous rock.

Post Road to Apulia.

  * N. B. A post is between 6 or 7 miles, but the Neapolitan mile is about 1¼ English.

  † A Diligence twice a week from Naples to Foggia.

Th. Duvdere delin.

J.W.Lowry sculp.

TOWN & CASTLE OF MELFI WITH Mᵗ VULTUR BEYOND.

SEEN FROM THE N.E.

Th.Ducros, delin.   ᵗ Rock of Melfi.   ᵗᵗ Mount Thabor.   a.a.a. Compact Lava.   b.b.b. Volcanic Tuff.   J.W.Lowry, sculp.

VIEW OF THE TOWN & CASTLE OF MELFI.

Th. Duclère, delin.

J.W.Lowry, sculp.

CRATER OF MOUNT VULTUR FROM THE S. W.

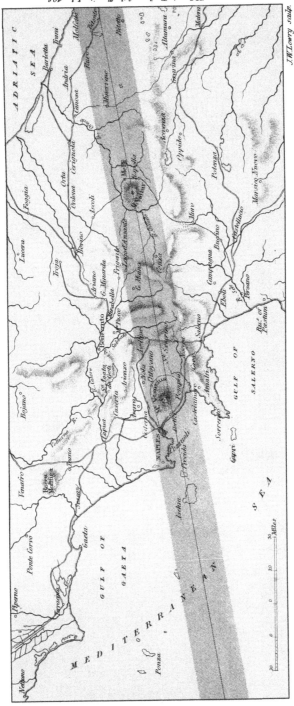

J.W.Lowry sculp.

MAP OF A PORTION OF THE KINGDOM OF NAPLES.

Printed in the United States
By Bookmasters